UNITY LIBRARY & ARCHIVES
The organic living book.
S 605.5 .K64 1972

0 0051 0043748 0

D1573141

The Organic Living Book

Also by Bernice Kohn:

The Beachcomber's Book

The Organic Living Book

BERNICE KOHN

Drawings by Betty Fraser

THE VIKING PRESS   NEW YORK

First Edition

Copyright © 1972 by Bernice Kohn. All rights reserved.
First published in 1972 by The Viking Press, Inc.
625 Madison Avenue, New York, N.Y. 10022
Published simultaneously in Canada by
The Macmillan Company of Canada Limited
Library of Congress catalog card number: 78-183936.
Printed in U.S.A.
1 2 3 4 5 76 75 74 73 72

574.5   Ecology
SBN 670-52833-1

For Morton,
who restored the natural order

CONTENTS

Introduction / 9
Learn to Be a Label-Looker / 12
You Just Might Learn to Love Yogurt / 19
The Organic Garden / 27
The Gardenless Gardener / 37
What Do You Do About Bugs? / 46
How to Sprout Sprouts—
 or the Twelve Inch Farm / 51
The Back-to-Nature Foods / 56
The Ceremony of Baking Bread / 62
The Organic Gourmet / 70
The Organic Ecologist / 76
For Further Reading / 85
Index / 89

INTRODUCTION

The title of this book may puzzle you somewhat. *Organic* is a common enough word, but if you look it up in the dictionary, you may be even more puzzled by the title. My dictionary gives fifteen definitions for the word organic, but none of them really explains what organic living is or what this book is all about.

To set the matter straight, then, this is a book about a good way to live. A way to live that is good for the whole earth, for the part of it that you inhabit, for you. It is a book about how to use the earth fully without using it up. About making the ground, the air, the water, and your body clean and healthy and strong and beautiful. It is about the rhythm of life and how to be part of the rhythm. About insects and animals and plants and people and all nature. How all living things live together in a natural order and how you fit into the order when you live in a natural way. It's a book about loving, too—about loving everything that's alive and caring for it. It's about making instead of destroying, about saving instead of wasting, about giving back to the earth as much as we take. This is a book about a good way to live, a book about organic living.

If you take those ideas one at a time, you have heard them all before. Using the earth without using it up is called *conservation*. Keeping the earth, air, and water clean is called *pollution control*. The study of how plants and animals live together in a natural way is called *ecology*. Organic living is part of conservation, pollution control, and ecology—but it is the part that is on the most personal, intimate level. It's the way you live every day.

Now, for the first time in the history of our country, many young people are fleeing our cities instead of flowing toward them. They are frightened by the noise, the smog, by too much concrete, too much crowding, by violence. They want to live and raise the children they hope to have where the air smells good and the flowers grow. They want to stick their fingers into the earth, to feel it, to *know* it. They want to grow wholesome food in the earth and eat the food.

Young people want to be warmed by a sun that doesn't have to push its rays through a pall of smoke. They want enough quiet to hear the birds sing. They have had too much of machines, of litter, of waste. They want to leave the cities so that they can feel the rhythm of nature again. This is the rhythm that their parents, in the quest for "progress," almost forgot, very nearly destroyed. For it is possible that with some ancient wisdom, long unused but deep in our cells, we may be coming to a halt just as we are about to kill the earth.

Not every young person can live on the land. Not every young person wants to. Many, for reasons of preference, or professional, family, or other ties, will never leave the city. But the good thing about organic living is that it doesn't matter where you are. Whether you live in a commune, on a ranch, on a farm, in the suburbs, in a house, an apartment, or a room in a ghetto, you can live better than you ever have before. You can make the whole world more beautiful. You can make yourself more beautiful, too. And perhaps most important of all, you can do it with love—and love doing it.

LEARN TO BE A LABEL-LOOKER

Everybody goes to the grocery store. Even if you live on the biggest farm in the world, you have to buy some foods at the market. The majority of us buy most of our food, so our first step in organic living begins in the grocery store.

The most wholesome, the most nutritious, and surely the best-tasting foods we can eat are organic foods. These are foods that have been raised and prepared without the use of poisonous sprays or chemicals of any kind. They are available from the many organic farms that are springing up all over the country and from the health food stores that can be found in every city and many towns. Supermarkets are beginning to carry organic foods, too, and some department stores have added organic food departments to their gourmet shops.

It is possible, however, that you do not live near any source of organic food, that you do not want to spend the extra money that unfortunately is called for, or that you are really not that much interested in pure organic eating. In that case, please don't close this book. Do your shopping in your neighborhood store or supermarket just as you always did—but do it with one difference. Read labels.

Most of us are creatures of habit. We like the foods we are used to, and since we recognize them at sight, we have little reason to read the label. If we do, we hardly ever read the fine print—unless we are interested in organic living. Organic-livers look at labels.

It is generally thought that if food is sold in a store, it must be fit to eat. Nothing is further from the truth. Every few years some widely used food is suddenly withdrawn from the market because it has just been found to be unsafe. Cases in point are the many drinks and other foods that were sweetened with cyclamates and the kinds of fish that were found to contain dangerous amounts of mercury. But many, many other foods are sold (and eaten) every day even though they contain harmful chemicals. Some of the chemicals are *known* to be harmful, but they are not considered poisonous in small amounts. Other chemicals have not been proven to be harmful—but they haven't been proven to be safe, either. Oddly—but luckily—for almost every food that contains a dangerous (or *possibly* dangerous) chemical, there is another brand on the market that is chemical free. All you have to do is read the label and choose the safe one.

The simplest rule to follow is to avoid anything that says "artificial," such as artificial flavor or artificial color; anything added to prevent spoiling or to preserve freshness; anything that has a chemical name instead of a food name. You might make some mistakes on that last one, but they can only be mistakes on the safe side.

Fortunately, manufacturers are finally becoming aware of the public's suspicion of food *additives*, as they are called. Many products ranging from soft drinks and ice cream to such unglamorous foods as cottage cheese bear the legends "no artificial flavor or color" or "no preservatives." Buy these brands in preference to others that contain chemicals.

It is a fact that the chemicals are usually used in tiny amounts, but there are hundreds—or thousands—of them in packages on the grocery shelf. Many of the tiny amounts of additives do not leave the body promptly but remain behind to accumulate into large amounts. In any ordinary day you very possibly might eat all the following, unappealing things.

In bread: sodium diacetate, chloromine T, potassium bromate, and calcium propionate.

In cake: sodium alginate, butyric acid, artificial flavors, and aluminum chloride.

In cereal: butyrated hydroxyanisole and sodium acetate.

In processed cheese: calcium citrate, sodium phosphate, aluminum potassium sulfate, and yellow dye.

In frankfurters and other meats: sodium nitrate,

sodium nitrite, asafoetida, magnesium carbonate, sodium ascorbate, and monosodium glutamate.

In soft drinks, fruit drinks, maraschino cherries, and many, many other products: benzoate of soda.

The paragraph that follows looks like something taken from a science fiction story, but it is actually the ingredient statement (the fine print) taken from the label of a well-known and widely sold whipped dessert topping that looks like whipped cream but isn't.

"Contains no milk or fat. A pasteurized blend of water, hydrogenated vegetable oil, sugar, propylene glycol monostearate, sodium caseinate, flavoring, potassium phosphate, salt, polysorbate 60, monoglycerides, sorbitan monostearate, lecithin, cellulose gum, carob bean gum, carrageenan, artificial coloring. Propellents: nitrous oxide, chloropentafluoroethane, carbon dioxide."

A popular diet soft drink contains carbonated water, citric acid, saccharin, sodium citrate, gum arabic, natural and artificial flavorings, brominated vegetable oil, salt, artificial coloring, and 1/20 of 1% benzoate soda and stannous chloride as preservatives.

Do you find it hard to believe that you actually *eat* such things? Read the labels in your own kitchen. How can you get food without additives? Easily. Next time, read the labels in the store instead of in the kitchen. You may have to do without a few old favorites altogether, but not many. In any ordinary market it should be possible to buy bread, cake, cereal, cheese, meat, and soft drinks, all without additives. You will find that in addition to being more nutritious, they all taste better.

No "dessert topping" can compete with real whipped cream!

If it seems like a lot of trouble to check labels, remember that you won't have to do it all the time. You will soon know which brands to buy. Remember, too, that we Americans have done a rather thorough job of polluting our air, our rivers, and our streams. It was done because it was "a lot of trouble" not to. Don't pollute your own body unnecessarily.

So far we have been reading labels to find out what has been added to food. You can find out from the same label what has been taken out—and most of the time, you won't even need the fine print.

It is one of the oddities of "progress" that many of our foods are so refined that most of the valuable food nutrients have been removed. White rice is not nearly so good for you as the brown rice it is made from. Brown sugar is better, honey better yet, than any refined sugar. Bleached white flour has had the best part of the wheat, the vitamin E, removed. Unbleached white flour, available every place, is better, whole wheat or other whole grain flour, better still.

It is another oddity that often almost all the nutrients are taken out and then the food is "enriched" by putting back some vitamins or minerals. Enriched, refined foods are *never* as good or as nutritious as the natural food is in the first place. Vitamins are used in complex ways by the body, and it is still not known whether man-made vitamins perform these complex functions as well as natural vitamins do.

Perhaps you are a little bit lazy about analyzing the

food you buy. Or you just don't want to give up certain foods. Or you are afraid that you will become a "food freak" or a "health nut." If so, just watch out for additives. Only that.

If you are willing to go a little further because you want to be as healthy as you can, to be as handsome as your natural gifts allow, and to avoid the wasting and contamination of what nature provides, you have to avoid more than additives. Whenever possible, do not use bleached white flour or anything made from it, white sugar, and white rice; hydrogenated oils, fats, and peanut butter. Eat all the peanut butter you want, but buy the natural kind. Better to take the trouble to stir it than to eat the chemicals that keep it smooth. They have never been proven safe for human consumption, and most authorities believe that hydrogenated fats raise the blood cholesterol level. Avoid cake mix, sugared dry cereals, and all other such "convenience" products. Try to use instead whole wheat, rye, or other whole grain bread or flour, raw sugar or honey, only liquid oil (without additives) or real butter for cooking and eating, non-hydrogenated peanut or other nut butters. Eat fresh rather than prepared meats (such as so-called luncheon meats like bologna), fresh fruits and vegetables as often as you can. And remember never, never to eat artificial colors and flavors or preservatives.

None of this makes you a "food nut." Looking at labels to know what you are buying is simple common sense. And it is the very first step on the road to organic living.

YOU JUST MIGHT LEARN TO LOVE YOGURT

You *might* love yogurt already, but many people don't. It's one of those foods that usually take some getting used to, but as soon as you *do* get used to it, you will more than likely become addicted. Especially if you make your own yogurt.

The main reason for making yogurt is that it's fun. It makes you feel like a cross between an alchemist and Louis Pasteur. Some other good reasons are that it tastes better than store-bought yogurt, costs much less, and is always fresh.

Of course it would be quite reasonable to ask why you should learn to love yogurt in the first place. There are some good reasons for that, too.

It's very, very good for you. It has all the nutrients of milk, but many of them are now partly broken down chemically and more available to your body. Yogurt is full of special bacteria that do all sorts of marvelous things for your digestive processes. It's low in calories, tastes good, and makes a great quick lunch or light dessert; it can be used in preparing many interesting foods and even be frozen with mashed fresh fruit to make a marvelous dietetic ice cream.

Since ancient times yogurt has been a staple food in the Middle East, southeastern Europe, and parts of Asia. You can find it as an ingredient of many of the specialties in any Greek, Armenian, Syrian, or Indian restaurant.

Throughout the Middle East yogurt is often eaten either plain or with honey or fruit as a dessert. It is also diluted with water and drunk icy cold as a refreshing beverage.

In countries where it is consumed in large quantities, it is common practice for a housewife to make yogurt in a big earthenware crock from which it is spooned out as needed. My own preference is to make it in one-portion containers since it tends to separate and get watery on top once some of it has been used. This does not affect the quality of the yogurt in any way since it is easy to stir it up with a spoon and mix it together again, but to me, it is never quite as appealing after it has lost its glossy, smooth top.

The basic step in yogurt manufacture involves adding bacterial yogurt culture to milk, providing the proper conditions, and then waiting for the bacteria to

multiply. There are almost as many ways of doing this as there are yogurt preparers. Each yogurt-maker undoubtedly has reasons for thinking his method the best, and I am no exception. Mine seems to me the simplest way, it is inexpensive, and it produces a yogurt which is both delicious and low in calories. Here then, is my favorite recipe for it.

HOMEMADE YOGURT

You will need (for 4 servings):

1-1/3 cups (or one 1-quart packet) instant nonfat dry milk powder
1/2 cup whole milk
A teakettleful of warm water
4 jelly glasses, or any glasses or cups with very thick bottoms
A heavy soup kettle or a casserole, with cover, large enough to hold the four glasses
2 heavy Turkish towels
A large mixing bowl
A spoon
1 container plain (unflavored) yogurt from the store. (Read the label and buy a brand without additives!)

Put half the yogurt into the mixing bowl and mix it with the half cup of whole milk. Stir it until it is quite smooth. (You can use an egg beater if you find it easier.) Add 2-1/2 cups of warm (not hot!) water. The temperature of the water is very important. If it is too cool, the bacteria won't multiply. If the temperature is too high, the bacteria will be killed. The water must be just comfortable to the touch, like a baby's bath water. If you are totally scientific and want to use a thermometer, the water should be between 110–120°F.

Add the dry milk powder to the mixture and stir until it is dissolved.

Stand the *clean* jelly glasses in the soup kettle or casserole and fill them with the mixture. Cover the glasses with their own metal tops, or if they have none, with plastic coffee-can covers or aluminum foil. Now pour some more of that nice bath-temperature water into the kettle until it is almost up to the tops of the jars. Cover the kettle and put the towels over the whole thing to keep in the heat.

You will have to check the temperature every hour or so. If it begins to feel cool, dip out one or two cupfuls and add some hot water to heat it up.

You can eliminate that last step entirely if you have a gas oven with a pilot light. Stand the kettle in the oven (without lighting it), and it should keep the temperature just about right. You can also stand the kettle on the pilot light of the stove, but this often gets too hot. Try yours, and if the water in the kettle starts to get too warm, put a hot pad over the pilot light. If you use either the oven or the stove, do *not* use the towels. They are unnecessary, and they can be a fire hazard.

Check the yogurt after 3 hours and see whether it is beginning to thicken. The yogurt is done when it is as thick as custard or soft ice cream—or when it looks like commercial yogurt. As soon as it reaches that stage, remove it from the kettle and put it into the refrigerator.

It almost always takes my yogurt about five hours to get done, but since there are so many variations in temperature and in the strength of the starter culture, yours may take more or less time than that. Usually, if the yogurt has not set within six hours, something is wrong. Either you did not have enough heat, or the culture was no good.

After you make your first batch of yogurt, save four tablespoonfuls to start your next batch. Try to do it within five days because after that time the bacteria become weak. For the same reason, your first container of "starter" yogurt must be fresh. Ask your grocer for the freshest container he has. Better yet, ask him when the yogurt truck comes to deliver and go to the store at that time.

Even though you will be using homemade yogurt for starter after the first time, you will have to buy a container every few months since the bacteria strain seems to weaken after a while. Whenever you notice that your yogurt is getting thinner or taking longer to set, buy a new starter.

The finished product should be absolutely smooth and have a mild flavor. It is normal for yogurt to be slightly tart, but if it has a really strong, sour taste, it was left in the heat too long. Also, too much heat can spoil the texture by making it tough or curdled.

LEARN TO LOVE YOGURT / 25

You don't have to make yogurt from powdered milk. You can use ordinary milk, which makes a much richer yogurt. The disadvantage is that you have to boil the milk first and then let it cool to lukewarm before you begin. Dry milk has already been boiled as part of the

evaporation process. However, if you like the richer yogurt—or are trying to gain weight—it is worth the extra step.

You can eliminate all the guesswork about temperature by buying an electric yogurt-maker. It has a thermostatically controlled heating element so that the temperature is always just right. This gadget also comes with its own flat-bottomed heavy glass dishes, complete with covers, and it is not necessary to immerse them in water. The only disadvantage is the initial cost, but if you plan to make yogurt often, you may find it worth the investment.

If money is no object, you might also like to start your yogurt with Bulgarian yogurt culture, which is available at health food stores. It is both pure and excellent, but the cost is quite high, and you have to buy a new batch every month.

A somewhat cheaper way to get an interesting culture is possible only if you happen to live near a Middle Eastern restaurant. If they have good homemade yogurt on the menu (and surely they do), see if they will sell you a yogurt "to go."

Eat your yogurt plain or with fresh fruit if you prefer it that way. If you like flavored yogurt, mix it with a spoonful of your favorite jam or preserves, with a little honey, or with honey and vanilla.

If you are a calorie watcher, use yogurt as a replacement for sour cream in all recipes such as salad dressings, cakes, or beef Stroganoff. Some yogurt recipes appear later in this book.

THE ORGANIC GARDEN

There is nothing very complicated about the principle of organic gardening. On the contrary, it is the simplest, most natural kind of gardening there is. It is the way nature gardens—without waste, without destruction, without poisons that upset the flow of life by killing birds, animals, or people. *Un*organic gardening is modern man's invention. Surely, the Garden of Eden was organic!

In the wild state, fruits fall to the ground and decay. Or they are eaten by birds and other animals, which then fertilize the ground with their droppings. Annual plants die at the end of their growing season and decay back into the earth, too. Fallen leaves and pine needles cover the ground with a thick blanket of mulch. Earthworms live in the rich soil and constantly plow it from within, filling it with air and worm castings, and this makes the earth soft and loamy.

Such beneficial creatures as ladybirds (ladybugs) and praying mantises abound where they have not been killed off by poison sprays that were intended for plant-eating insects. Unfortunately, sprays do not make distinctions between useful and harmful insects. But helpful insects and birds do, so where they are allowed to live, they do the "insecticide" job in the most natural way there is. At the same time, they are assured of an adequate natural food supply so that they will be back next year and every year after that, too.

When an organic garden is not wild but cultivated, the idea is to do everything as close to nature's way as possible. Fertilize the earth with the very things that grew out of it. Enrich it further with the waste products of animals. When harmful insects have to be killed, let their natural enemies do it.

There are innumerable books on organic gardening available. If you plan to garden on a large scale, you will want as much information as you can get. There are several excellent reference books listed in the bibliography at the end of this book. If, on the other hand, as a beginner you want to start on a small scale or even just dabble at first, there are only a few things you absolutely must know.

The basic plan is to use the earth without using it up. Put back everything you don't need. There is more than enough for you and *it*. Think ecology. The plants, the birds, the insects, the animals, and people all work together in a natural way. Don't add anything to your garden that can harm any living thing. Don't add anything artificial. Enrich the soil with natural fertilizers, and you will be astounded to see what comes out of it. Whatever you grow will be stronger, healthier, larger, more fragrant and will taste better than any other kind of garden produce there is. The way to achieve this small miracle is with the compost heap.

There is really nothing magic about compost. It is simply an easy way of *recycling* (using over, or conserving) our waste and putting it back to work in the garden. Instead of filling up smelly garbage cans or polluting the air by burning leaves or forcing people to pick their way through city streets littered with animal manure or polluting our rivers with sewage, we use all of these organic wastes to produce tomatoes and raspberries. If this were done universally, we would have far fewer pollution problems and much better food—but that may take a while. Let's just start with your own compost.

There is no one right way to make a compost heap. You can read about many ways, and if you try them, you will probably get similar results from all. The method that is described here is an easy way to make compost without being too scientific, industrious, or rich.

First of all, you need some kind of container. You can use a garbage can. If you have some space, make an enclosure of some sort on the ground, about three feet by three feet and perhaps four feet high. Use old boards, chicken wire, concrete blocks, anything at all that you can find that will form a fence. The only purpose of the enclosure is to keep the heap together so that it doesn't spread all over the place. It will be easier to work if you make the sides—or at least one side—removable.

The particular method of composting that follows is based on the work of Sir Albert Howard. Since his concept was developed in Indore, India, it is known as the Indore method.

It is essential for compost to have a flow of air under it and to be well drained, and so you must begin by putting coarse material on the bottom. A layer of heavy brush, limbs pruned from trees, or twigs are all fine for

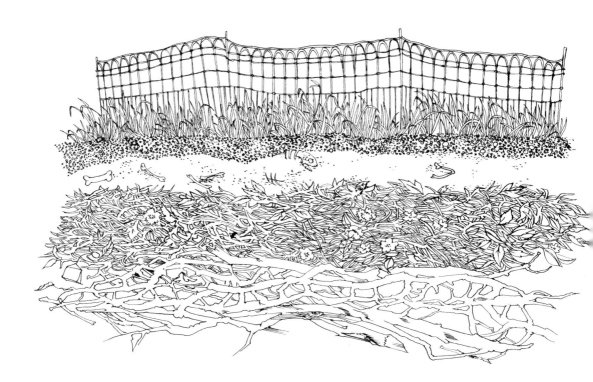

this purpose. If you don't have and can't find any of these, prop up a few boards or slats so that they are a couple of inches off the ground and crisscross a few others over them to make a latticed floor. Now start your composting layers.

Most of the heap will be made of *green matter*, which is a very broad term. It means just about anything that ever grew in the ground whether it is, ever was, or was never green. You can use grass clippings, old leaves, weeds, dead flowers, coffee grounds, tea leaves, leftover salad, carrot tops, banana peels, all the waste of kitchen or garden that will *decompose*. In kitchen garbage, you will of course have to keep separate—or remove—cans, bottles, and plastic items. Don't worry about the paper napkin that got mixed with the old spaghetti. Paper came from a tree and is perfectly good organic material.

After you have put down six inches of green material, you are ready for two inches of manure. This can be the fecal waste from any kind of animal (except human beings, which is illegal for use because of the possibility of spreading disease). Use manure from cows, horses, ducks, chickens, pigs, sheep, and so forth. If you do not live in the country and cannot obtain such manures, follow your neighbors' dogs—or offer to empty an apartment-dweller's cat pan for a few days.

If available, you can also toss in with the manure layer things like fish heads and chicken entrails or bones. The whole idea is to provide all the nutrients your garden needs plus a batch of bacteria that will break the compost down into good, rich *humus* that the garden can use. In addition, you have to provide moisture—and just a little bit of muscle.

Don't get a ruler to measure six inches of green matter and two inches of manure. Just remember that there should be about three times as much green matter as manure when you make the layers.

It doesn't make any difference where you make your compost heap because—and this may surprise you—you will find that it is completely without odor and not at all unsightly. As soon as you have finished building the compost layers, cover the top completely with earth and wet it down so that the whole pile becomes moist. Because of the rich material it contains, the top layer of earth will soon put forth a carpet of green weeds.

For a while, you will have nothing to do except water when necessary to keep the compost damp. If it rains, you don't even have to do that. The bacteria inside the pile do all the work. As they digest the material and turn it into humus, you will notice two things: the interior of the heap will rise in temperature until it is much warmer than the outside (stick in a finger and see), and later the top of the heap will sink down so that it is slightly concave.

You can help the composting process on its way by assuring a good supply of air. Take an old broomstick or a pole of some sort and poke a few holes here and there once in a while.

After six weeks, it is time to turn the compost. Remove the sides of the bin, or the one side, or shovel from your garbage can and then back into it again, or whatever, but turn it all over. After twelve weeks (six weeks after the first turn) turn it again, and now it is humus, ready for use.

Although you can start compost at any time, I like to start mine in the fall. I always have plenty of green matter at that time—all the old garden plants such as zucchini vines, tomato plants, dead leaves, and hay. Also, I am in no hurry for the compost to be finished since I have all winter to wait, and it does the heap no harm to sit. In actual practice, I usually have two heaps going—last year's in use during the spring and summer and the new one getting ready for next year.

If you use your compost after twelve weeks, don't expect that every cornstalk or orange peel will have broken down into a fine, black powder. It certainly will not have done so. There will be many large, recognizable chunks, some looking no different from the last time you saw them. Don't worry about it. If you are using the compost in a small planting area such as a flowerpot or box, just pick out the large pieces and throw them back in the heap. There will be enough fine particles full of rich nutrients without these.

The Indore method of making compost is a classic one, but for one reason or another you may not be able to follow it. Simplify if you have to, even invent your own method. One caution—you cannot make compost from a single ingredient. You can water a pile of grass clippings or leaves for a long, long time, and they may rot or turn moldy or do any one of a number of other things, but they won't turn into compost. You must provide a rich source of nitrogen and the manure is included in the Indore "recipe" for that reason. If you

can't get any manure, find another nitrogen source. Get some bloody flesh and bone scraps from your butcher or fish dealer, the sewer sludge that is packaged and sold by an increasing number of towns and cities concerned with ecology, or just a sackful of good, rich forest earth.

If they are available to you, some other good materials to throw into the compost are sawdust, wood chips, nutshells, and packaged dried blood from slaughterhouses.

Just remember not to include anything that is too large because it will take it a long time to disintegrate. The smaller the material you start with, the faster the humus will form. Some gardeners use special shredding machines, and with one of these it is possible to make compost in only fourteen days. If you have no shredder, cut up large items such as melon or grapefruit rinds before adding them to the compost heap.

Once your compost is finished, you have to know what to do with it. Compost is food for your plants, and the *only* food or fertilizer they will ever need. Sprinkle it on your lawn and add a two- or three-inch layer to your garden once a year. You can use it when you turn over the earth and get ready to plant in the spring. Just work the compost into the top few inches of earth. Or in a garden that is already growing, mix the compost with some topsoil and spread it between the plants. This is called *top dressing* or *side dressing*.

After your plants are growing well and are at least several inches tall, spread a mulch—a few inches of straw, hay, grass clippings, sawdust, or any other

organic material—between the rows and around the plants. Mulch serves several purposes: it keeps in moisture so that plants don't dry out quickly in hot weather, it helps keep down weeds, and it decomposes slowly and adds its nutrients to the soil. In the spring, when you get ready to plant your new crops, work the old mulch right back into the earth. Add your compost and you're all ready to go.

The ordinary aspects of gardening—weeding, watering, cultivating, and so forth—are the same for the organic garden as for any other kind. The only big difference is in pest control. Although that was touched on briefly at the start of this chapter, it is explained in more detail later in this book.

One last word: If you have a yard but don't want to garden or be bothered with compost, recycle your garbage anyway. Dig a hole and bury it and give it back to the earth.

THE GARDENLESS GARDENER

The organic gardening bug has bitten you. You are sure that both your thumbs are green—but you would like to prove it. You are quite ready to believe that nothing tastes like a freshly picked vegetable—but you have never tasted one. You would love to eat a tomato that has never been sprayed with poison or an unwaxed cucumber that you don't have to peel or a strawberry so pure that you can eat it right off the plant. In short, you want a garden. But you have no real estate.

Take heart! There is a way. In fact, there are many ways. Just pick out the ones that suit your needs. If you have a porch, a balcony, or a terrace, you can

garden in pots, baskets, or boxes. You can do the same thing (with the landlord's permission) on an apartment house roof. If you have no outdoor space of any kind, you can be an indoor gardener. If you have a sunny window, you have no problem. If you have no sun at all, you can solve that problem with a special kind of electric lamp that you can purchase from a garden supply store. And if you have no window and no electricity, you probably aren't reading this book.

Whatever your locale, let your imagination run riot for plant containers. Depending on the amount of space you have, use flowerpots, flower boxes, a bushel basket, a plastic laundry basket, dishpan, barrel, scrub pail,

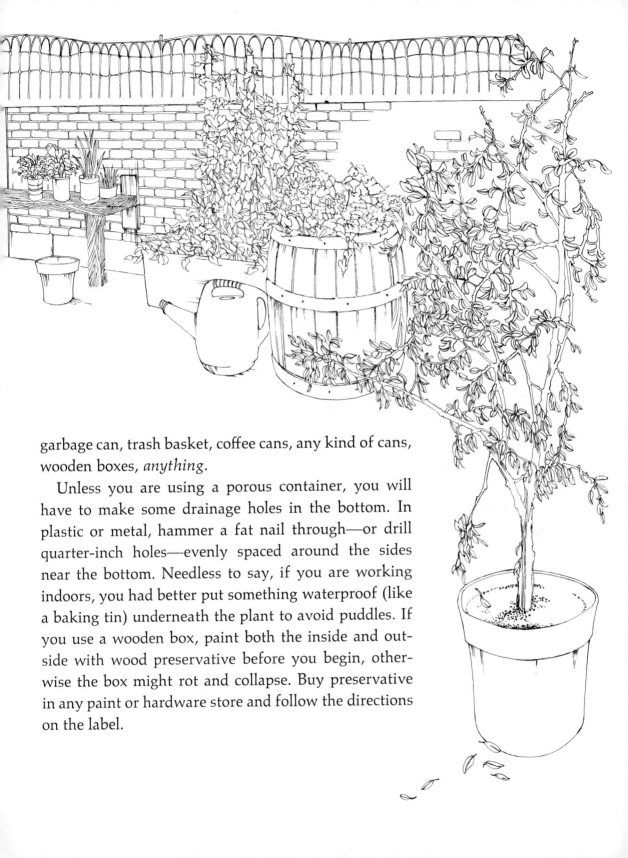

garbage can, trash basket, coffee cans, any kind of cans, wooden boxes, *anything*.

Unless you are using a porous container, you will have to make some drainage holes in the bottom. In plastic or metal, hammer a fat nail through—or drill quarter-inch holes—evenly spaced around the sides near the bottom. Needless to say, if you are working indoors, you had better put something waterproof (like a baking tin) underneath the plant to avoid puddles. If you use a wooden box, paint both the inside and outside with wood preservative before you begin, otherwise the box might rot and collapse. Buy preservative in any paint or hardware store and follow the directions on the label.

When you are ready to plant, put some gravel or a few pieces of a broken flowerpot or pottery in the bottom of the container so that the earth doesn't cake down and prevent good drainage. Fill the container with some rich garden earth from outdoors if you can get any. If not, buy potting soil at your local nursery or florist shop.

If you are going to plant seeds, you can start them right in the main container, or you can start them in special seedling containers first. If you are going to use an outdoor planter, the latter is the best plan because then, unless you live where it is never cold, you can get your seeds started early and ready to put outside as soon as the weather is warm enough.

Plant seeds in either soil or vermiculite (a dried, granular mineral preparation), in peat pots, plastic bags, styrofoam coffee cups, cut-down milk cartons, tin cans, cigar boxes, or cake pans. One good trick is to use a coffee can with both ends removed and the plastic cover put on the bottom. When it is time to replant, it is easy to take off the bottom and push out the plant without doing any damage to the plant or its roots.

All green plants need light, but some need more than others. The amount of sunlight you have in your planting area will help you decide what to grow. Most of the leafy green vegetables need plenty of sunlight. So do cucumbers, eggplant, peppers, summer squash, and tomatoes. Vegetables that do well in partial shade are beets, carrots, cabbage, chives, kale, some varieties of lettuce, parsley, radishes, and scallions.

Many herbs do well with little sun. Since many of

Thyme

these are not commonly available in stores in their fresh state, they are a particularly good choice for the gardenless gardener. You might try lovage, sweet marjoram, lemon thyme, basil, tarragon, or mint.

If you have no more space than a window sill, try a small selection of herbs that are quite different from each other. Such an assortment might be parsley, chives, tarragon, and garlic (plant a few cloves of your kitchen garlic and clip the green tops as they grow for salad or cooking).

Basil

If you have room near a window for a good-sized flower box or some container as large as a bushel basket, you can raise an entire salad. Plant Tiny Tim tomatoes, radishes, one or more of the leaf lettuces—salad bowl, buttercrunch, ruby, or bronze—and upland cress. You can also plant one of the short, stubby varieties of carrots, and if there is room, a few nasturtiums. The flowers are bright orange and yellow, and in addition to their being lovely to look at either in the planter or in a bouquet, a few of the blooms can be tossed into your salad bowl. They are guaranteed to attract attention—and they taste good. The stems and leaves of the nasturtium can be a good addition to salad too, but they are very peppery, so chop them well and use them sparingly.

Tarragon

As soon as you have made your decisions about crops and prepared your planters, you need something to plant. The easy way is to go to a store or open a mail-order seed catalogue and buy some packets of seeds. It may also be the *best* way because seeds that are old often do not grow. Make sure that any seed packets you buy have the current year's date on them.

But instead of or in addition to bought seeds, you can have some fun by experimenting to see what you can grow without buying anything. Try some from the jars on your pantry shelf—but be certain that they are seeds, not crushed leaves. Some of them might be dill, caraway, coriander, or cumin. Soak some dried beans or peas for a day or two and plant them. Plant orange, lemon, or grapefruit seeds. Some of these things might grow, some might not. In any case, it's fun to try, and you have nothing to lose.

If you buy seeds, you will find planting directions on the packet. If you are planting outside, you have to be sure that it is not too cold. A good rule to follow in northern climes is to plant when the oak leaves are fully out.

Plant a few seeds at a time, and when the seedlings come up, discard any that look weak and straggly and thin out the others to the distance recommended on the package—or if you have no package, plant them several inches apart.

It is important to water your plants regularly to keep them moist. Watering doesn't mean drowning, though. Too much water will kill your garden just as surely as too little. A pretty good soaking two or three times a week should be enough. It's all right for the earth to look dry on top as long as it feels moist when you stick your fingertip into it.

If your garden is indoors and will remain indoors, you are on your way. If, however, you have started seeds in the house and plan to transfer the young plants outside, you will have to harden them first. For a period

of two weeks, give them much less water and put them near an open door or window for several hours a day or move them to a cooler room. Hardening makes the plants less tender and will keep them from being killed by the wind and by too high or too low temperatures.

Be sure to weed your garden. Weeds rob plants of water, space, and nutrients and, if they are allowed to grow tall, of light as well.

You will need compost for fertilizing, and a covered garbage pail is just about the right size compost container for the gardenless gardener. And how appropriate it is! A garbage pail full of compost is almost a symbol of the organic idea: Keep the life cycle going by returning the fruits of the earth to the earth.

A final suggestion for the northern gardenless gardener who can't wait until spring to plant: you might just have to wait for most things, but if you want to try some rather unusual salads, make an indoor winter salad garden. Buy some root vegetables—carrots, turnips, parsnips, beets, celery root, whatever your market happens to have. Plant them pointed end down in a fairly deep (at least one foot) container that is filled with sand. *Just* sand. No soil, no compost. Keep the sand moist, and soon the buried roots should begin to send up green sprouts. While they are still young and tender, snip them and make your winter salad. You might have several containers going at one time. Plant them at one- or two-week intervals and keep yourself in salad greens all winter long.

If you raised any herbs, you can have another kind of winter-long garden pleasure. Almost all the herbs can be dried successfully. Pick them when they are in their prime, while they are flowering, or immediately afterward. Break off the sprigs, rinse them under running water, and shake them well. Now spread the herbs on a towel on a tray or a cookie sheet. Put them in a warm dry place—an attic is fine if you have one—but keep them out of the sun. Direct sunlight changes the green herb color to a muddy brown.

Turn the herbs over once a day until they are thoroughly dry or will crumble easily between your fingers. You can store them in two ways. The usual way is to crumble them up, stems and all, and put them into airtight jars. Small baby food jars are excellent for this purpose. Label each jar so you don't forget what it is.

My favorite way of jarring dried herbs is to put the entire sprigs into jars (large baby food jars this time) complete with flowers if there are any. They look extremely attractive, and I think they keep their aroma a little longer in the whole state. It's easy enough to crumble a pinch when needed.

Small bouquets of various herbs tied with a ribbon make charming gifts.

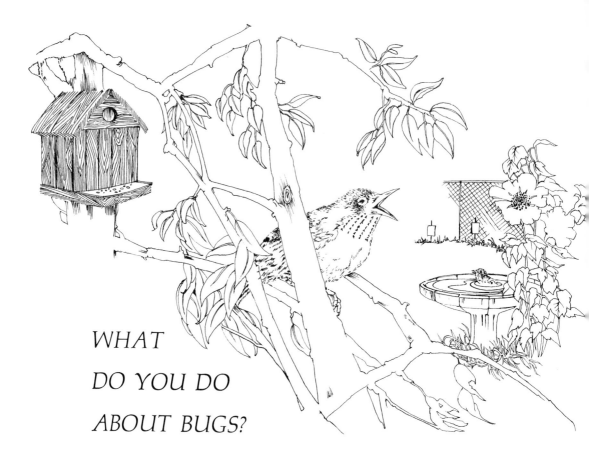

WHAT DO YOU DO ABOUT BUGS?

In the year 1962, Rachel Carson's book *Silent Spring* was published. The book created a furor that was nothing short of extraordinary, considering its subject matter. Miss Carson pointed out—most eloquently—that the pesticides that were then being widely used were doing much more than killing insect pests. They were interfering with the natural order of things and upsetting the ecological balance. Pesticides were, in fact, polluting the earth. Poisons not only killed birds directly, but frequently starved them to death by eliminating the insects upon which they fed. The book warned that unless the use of certain poisons were

WHAT DO YOU DO ABOUT BUGS? / 47

banned, we would all awaken one spring day to a world without birdsong, a silent spring.

The furor arose because it was clear that if birds were first, human beings were next. Every living thing is connected to every other living thing in the ecological chain. The death of any plant or animal will affect the entire chain. The prospect of a spring without birds was so terrifying that the simplest way for many persons to deal with the prediction was to deny its truth. The controversy raged between those who believed in Rachel Carson and those who believed in DDT.

For the first time, government and citizens became publicly and seriously concerned with what we were doing to the earth. Investigations began, and some regulations went into effect. Much good was accomplished. Much remains to be done.

The organic gardeners knew all about pesticides and ecology long before *Silent Spring* was published. They were far ahead of the general public and far ahead of government regulation. They are still ahead today.

While we have banned the use of many dangerous poisons and control the use of others, the organic gardener doesn't use them at all. Not only because he believes that they are harmful, but because he doesn't need them.

Now it is true that an organic garden has a variety of insects crawling, flying, and digging through it. But two other things are also true: most of the insects are harmless to plants—if not positively beneficial—and the insects that *are* harmful are much less apt to attack healthy, strong, organically grown plants than any

48 / THE ORGANIC LIVING BOOK

other kind. This does not mean that no beetle will ever nibble your beans, no cutworm will ever kill a tomato plant. It does mean that you can maintain a garden without the use of any poisons at all and harvest a bigger and better crop than the people who do.

Nature has its own way of balancing things. In virgin forests, where no one has ever sprayed, the trees flourish and so do the flowers and ferns. Good insects eat bad insects, birds eat both kinds, and it works out. There is enough for all, and the cycle runs its course.

The same thing can be true in your garden. Some of your plants *will* be attacked. But if you are willing to share with the insects, you will still have enough. The damage will be small. If you try to poison the bad insects, you will poison the good ones too, and very likely, the birds, the animals, and yourself as well.

Now, this doesn't mean that you can't use reasonable means to protect your plants. "Reasonable" means doing it in a natural way. Here are some natural ways to keep your garden growing without insecticides.

Keep your plants strong with good compost and proper light and water. Keep your garden well stocked

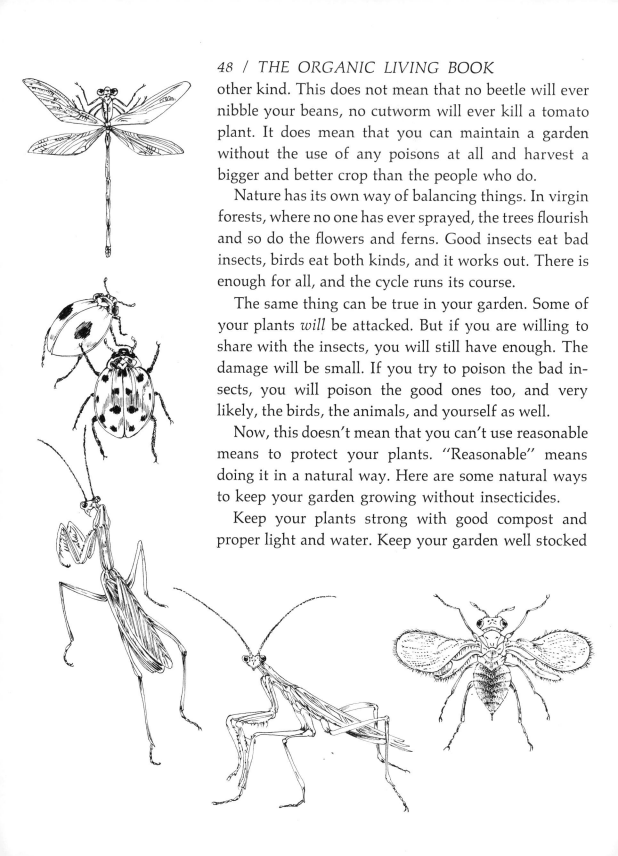

WHAT DO YOU DO ABOUT BUGS? / 49

with beneficial insects. If they don't live there already, buy some from an organic gardening supply house. They don't cost much. Get praying mantises, ladybirds, damselflies, trichogramma wasps, and tiger beetles.

Attract as many birds as possible. Build birdhouses, provide fresh water in a birdbath or a bowl. Put out a bird feeder or, better yet, cultivate some plants that attract birds. These include sunflowers, asters, cosmos, and almost all bushes that bear berries—honeysuckle, snowberry, barberry, mulberry.

One toad eats several thousand insects every month. Toads need water, so keep a few panfuls in your garden if there is no natural pool nearby. If you make the conditions attractive enough, toads that have become established on your territory will remain for generation after generation.

In addition to all these natural-enemy methods, there are a few tricks you can do all by yourself. Insects, like people, apparently have their dislikes as well as their likes. There are some things they just can't stand and will go to great lengths to avoid. Make the most of it.

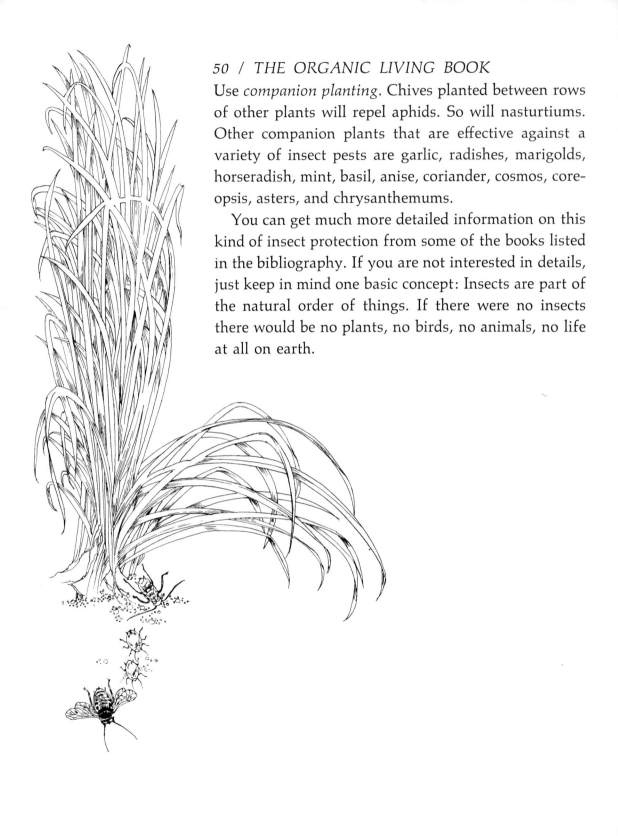

Use *companion planting*. Chives planted between rows of other plants will repel aphids. So will nasturtiums. Other companion plants that are effective against a variety of insect pests are garlic, radishes, marigolds, horseradish, mint, basil, anise, coriander, cosmos, coreopsis, asters, and chrysanthemums.

You can get much more detailed information on this kind of insect protection from some of the books listed in the bibliography. If you are not interested in details, just keep in mind one basic concept: Insects are part of the natural order of things. If there were no insects there would be no plants, no birds, no animals, no life at all on earth.

HOW TO SPROUT SPROUTS— OR THE TWELVE-INCH FARM

If you are even less of a gardener than the gardenless one, sprouts are absolutely the end of the line in mini-farming. You don't need earth, sand, compost, or sunlight. You don't even have to plant anything.

Have you ever eaten in a Chinese restaurant—and do you remember those crunchy little bean sprouts? Bean sprouts are beans that have *sprouted* or started to grow. Every bean stores enough nutriments from the parent plant so that even before it has acquired the root system that enables it to get new nourishment from the earth, it is equipped to form a new plant. That's where you come in.

Stored up in that compact little bean is a whole wealth of vitamins, minerals, proteins, and other food elements. And when the sprouts begin to grow, the vitamin content is said to be actually increased. Here is the perfect crop for the would-be gardener without a garden. Sprouts can be grown by anyone, anywhere. They require no land, no soil, not even a flowerpot, no light, almost no space. All you need is a colander or sieve, a bowl, some water, a towel, and something to sprout.

In case you need more to inspire you than the high nutritional value of sprouts, it should be mentioned that they are fantastically delicious to eat. Sprouts are crunchy and fresh-tasting. You can eat them raw or mix them with other foods in Chinese style. Or put them in muffin or pancake batter. Or in soup or stew. Once you have eaten a salad with sprouts, you will never want a sproutless salad again.

Sprouts can be grown from almost any kind of edible bean, pea, grain, or seed. These include lima beans, lentils, soybeans, mung beans, garbanzos (chick peas), fava beans, marrow beans, kidney beans; dried peas; wheat, rye, and barley grains; alfalfa, sesame, and sunflower seeds. If you can get organically grown beans or seeds, do so. If not, use ordinary ones from the market. Do not use packets of seeds intended for planting unless the label specifically says that the seeds have not been treated with any chemicals.

My own favorite sprouts are mung bean, and when I am unable to get to a health food store, I buy the beans in an oriental food store.

HOW TO SPROUT SPROUTS / 53

As with making yogurt, there are innumerable methods for sprouting, and they probably all work. This one works well for me, and it doesn't need any fancy equipment.

SPROUTS
You will need:
Beans or seeds
A bowl
A colander or sieve
A towel
Running water

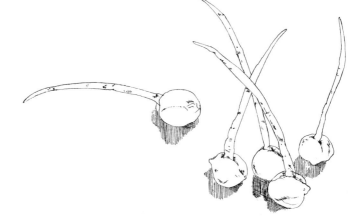

Use about 1/2 cup (or less) of beans, seeds or whatever. Keep in mind that when you are finished, you will have about six times the quantity you started with. Pick over the beans, discard any that are broken or spoiled, wash them, and soak them overnight in a bowl of water.

After soaking, pour the beans into a colander and spread them around in a single layer. If you are using tiny seeds that would fall through the holes, use a sieve instead. Or use the colander but put a piece of clean, thin cloth in the bottom first. Now stand the colander in a bowl and cover it with a towel or cloth to exclude the light. The whole idea is to keep the inside moist and dark, but airy.

Three or four times a day, uncover the colander, lift it out of the bowl and hold it under cool running water for a few seconds, rinsing all the beans thoroughly. The rinse serves two purposes: it keeps the beans moist,

54 / THE ORGANIC LIVING BOOK

and it flushes away fungi or bacteria that might cause spoiling.

The only spoilage problem I have ever had was with soybeans. They take several days to sprout, and most of the time they do it beautifully. But once in a while, especially in very hot weather, they ferment and I have to throw away the whole batch. If you wonder how you will recognize fermentation, trust your nose.

Most sprouts are ready in about three days. Soybean, mung bean, and pea sprouts should be about 2 inches long, alfalfa-seed sprouts between 1 and 2 inches, and most other seeds a little less than that. When your sprouts look about right, taste some. They should be crisp and crunchy, but the beans or seeds must be soft enough to eat. The sprouts should not be allowed to grow so long that they develop rootlets.

As soon as the sprouts are finished, give them one last rinse, shake out as much water as possible, and store them in a covered container in the refrigerator. They will remain fresh for about a week—if you can

keep from eating them that long. Some people freeze sprouts and keep them for months, but I have never had success with freezing them. My sprouts turn mushy when thawed and lose their flavor, too. You might try freezing some and see what you think.

Just for fun, stick a few sprouts (fresh, not frozen ones) into the garden or a flowerpot with the sprout tips just sticking out of the earth. Water them every day and see what happens.

Toss a handful of sprouts into your next batch of scrambled eggs or an omelet. Good.

THE BACK-TO-NATURE FOODS

The last few chapters may have given you the impression that you have to be a gardener, at least on a miniature scale, in order to live organically. Of course that is not the case. If growing it yourself is not your thing, you can get back to nature (once removed) by providing yourself with natural food that someone else grew for you.

Earlier I explained the simplest approach, that of avoiding certain foods. I also touched on the direct approach of seeking out natural foods. Now that you have more background, we can explore natural foods a little more thoroughly. The same thing can be said of natural foods that was said about organic gardens. They are foods that are just as nature made them, foods that have not been "improved" by man.

Natural foods include all whole grains that have not been *fumigated*—treated with chemical gases to prevent spoilage. Without exception, whole cereal grains are more nutritious than refined grains, which have been divested of the best part of their vitamin content. Natural grains include whole wheat, buckwheat, bulgar, rye, millet, brown rice, oats, cornmeal, bran, soy grits.

Flours should be made of whole grain, be as freshly ground as possible, and preferably, stone ground. Grinding with stones produces very little heat and does not destroy vitamins as conventional grinding does. Almost all flours sold in health food stores are stone-ground ones, and they can also be found on the shelves of some regular stores.

Butter should be fresh and without any added coloring. Watch out for margarines. They are usually hydrogenated and contain other additives as well.

All fats for cooking should be liquid, unhydrogenated, all-vegetable, and unrefined. Safflower, olive, corn, sunflower, peanut, soybean, and cottonseed oils are acceptable. But again, read the labels. They are full of surprises. A common brand of safflower oil (which most people think of as organic) sold in many markets contains a preservative. But the most popular (at least in my part of the country), brand of peanut oil, sold in every market, is one hundred per cent pure peanut oil and contains no additives at all.

Honey that is pure and untreated in any way is the best natural sweetener there is. In addition, you can use pure maple syrup, unsulphured pure molasses, or raw sugar. Raw sugar, also called turbinado sugar, can be used in your sugar bowl and for baking. In general, dark sugars contain minerals that have been removed from white sugars.

Cheese, ideally, should be made from whole raw milk or skim milk, but all cheeses are acceptable as long as they contain no additives and are not processed. Processed cheese contains hydrogenated fat. Read the label. If it says "process cheese," put it back.

Pure buttermilk, yogurt, dried unsulphured and unfumigated fruits, unsweetened fresh fruit juices, unsprayed fresh fruits, vegetables and herbs, plain gelatin, untreated dried beans, nuts that are raw and without added oil or salt (so that they retain their full value without additives), all are fine foods. So is meat from animals that have not been given hormones (if you can get it). Hormones, which make the animals grow bigger and fatter, remain in the meat and are very highly suspect of producing cancer in human beings.

If you want to go even further, buy certified milk and other raw milk products such as cheese and cream. Raw milk does not require vitamin-destroying pasteurization because it has been certified to come from cows that are free of disease. You can also buy fertilized eggs that come from organically raised chickens. Not only are such eggs more nutritious than ordinary eggs, but the chickens have not been fed on chemicals or additives. All these products must, of course, come from reputable dealers.

The most natural food of all, of course, is wild food.

Although I have a sizable organic garden and take great pleasure in eating my own tomatoes, squash, and raspberries, it is a completely different experience to find, identify, and eat wild food. It gives me a sense of the flow of time, with its thread of continuity from the beginning into eternity. Not only do I forage in precisely the same way as the Montauk Indians who once roamed Long Island, but it is entirely possible that I may even be following in the footsteps of some long-ago cavewoman.

Sorrel

A weed that grows in great profusion in my garden is purslane. All my neighbors consider purslane a pest, but a good bit of mine winds up in the salad bowl. It is crisp and succulent. Sheep sorrel, another abundant "pest," makes one of my favorite soups. Clams, crabs, mussels, and seaweed are free for the taking all along this miraculously still unpolluted seacoast. So are many kinds of fish. Eaten just minutes out of the water, they are unusually delicious and just as natural as can be.

Many people probably know about the joys of wild berries, but anyone who doesn't, should. Wild berries are generally smaller but much tastier than cultivated varieties. Strawberries, raspberries, blueberries, mulberries, blackberries are all fairly easy to find in many parts of the country.

There are a number of wild plants that make excellent beverages, too. My favorite among these is sassafras tea. Sassafras is a common plant that grows almost everywhere. The root is the part used to make tea, and sassafras is one of the few plants I would ever suggest pulling up by the roots. Its growth is such that it is al-

Purslane

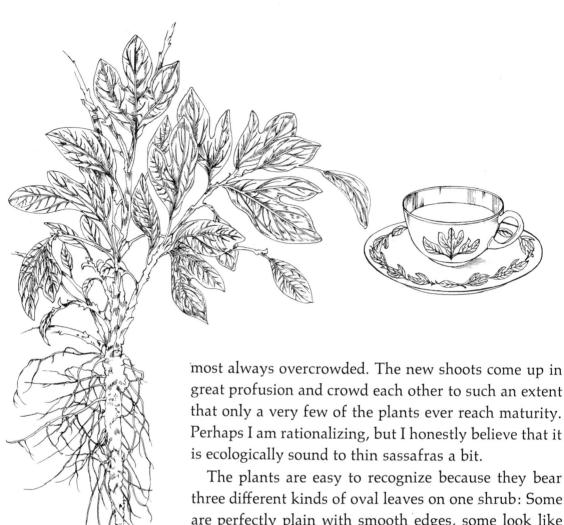

Sassafras

most always overcrowded. The new shoots come up in great profusion and crowd each other to such an extent that only a very few of the plants ever reach maturity. Perhaps I am rationalizing, but I honestly believe that it is ecologically sound to thin sassafras a bit.

The plants are easy to recognize because they bear three different kinds of oval leaves on one shrub: Some are perfectly plain with smooth edges, some look like mittens with a thumb on one side, and some look like mittens with a thumb on either side. While at maturity sassafras is actually a tree, the plants you will want for tea-making purposes will be little bushes about a foot high. When you think you have found a sassafras bush, pull it up carefully and smell the long root. A sassafras root smells like root beer.

After you have collected a few roots, wash them thoroughly and put them out to dry. Several days later,

THE BACK-TO-NATURE FOODS / 61

break or cut them into half-inch pieces and store them in a jar. Caution: If the roots are not completely dry, they will become moldy (as mine did the first time).

When you are ready to make tea, put about half a handful of roots into a saucepan with a couple of cups of water. Boil it for about ten minutes or until the water attains a lovely ruddy color and gives off a rich fragrance. Drink the tea with a little lemon and honey in it.

Sassafras tea has a long history in the United States. It was a beverage much used by the American Indians. When the colonists learned how to make it, they drank it as a substitute for the highly taxed British tea that was one of the points of contention that led to the War for Independence.

There are some books about wild food in the bibliography, and reading them is bound to enrich your organic life. Of course before you eat any wild food, it is important to be *absolutely certain* that you know what it is. And unless you are a world-renowned expert on them, *stay away from wild mushrooms*. One mistake can be fatal.

While poisonous mushrooms are an extreme case in point, they should make you think in terms of your own interior ecology. We are accustomed to thinking about the ecology of the earth and the creatures that inhabit the earth—but of course, you are one of those creatures. Just as every organism in the world is a part of the whole pattern and flow of life, so everything that goes into your body becomes a part of the flow of life within your body. You might even say that just as charity begins at home, your own ecology begins in your mouth.

THE CEREMONY OF BAKING BREAD

While it is possible to buy good organic bread almost everywhere, there is something very, very special about baking bread yourself. Homemade bread is certainly good to eat, but the product is only a part of the reason for baking. The doing is special, too.

Just as digging in the earth and growing food seem to satisfy a long-unsatisfied urge in many of us, so kneading a lump of dough, shaping it, and baking it into a crispy, fragrant loaf, takes us back to a time before our time. The baking of bread is in some way a magical, almost a religious rite.

It is not accidental that bread has been used as a sacrament since ancient times. In *The Golden Bough,*

Sir James Frazer points out that the Christian sacrament was adopted from a tradition much older than Christianity itself. In Asia, there was a very old sacrament involving millet bread among the Ainu of Japan. Frazer believes that although the homage to and the partaking of the millet cakes by the Ainu are tributes to a god, the god is really the grain itself because it is beneficial to the human body. While Christian custom did not derive directly from the Ainu, it was part of a long chain of cultures and religions that used bread in various sacramental ways.

In the New World, bread made of corn was eaten long ago by the Aztecs who believed it to be the body of a god. This custom prevailed until the conquest of Mexico by the Spaniards. The bread sacrament was, of course, introduced in new form when the Spaniards converted the Indians to Catholicism.

The importance of bread in our own not unfailingly reverent culture is made clear in such expressions as "to earn one's bread" for earning a livelihood; "bread" as a slang expression for money; "to break bread" to indicate the partaking or sharing of any kind of food. Obviously, bread is basic. Made in the old way, bread is basic, natural, and organic.

The simplest possible kind of bread is just a mixture of flour and water that has been baked. It does not have to rise. The *matzoh*, or unleavened bread, that was made by the Hebrews of the Bible when they fled Egypt was such a bread.

Many people bake unleavened breads today, and there are recipes for them in most natural food cookbooks. Most of us, however, are more accustomed to bread that contains yeast plus a few other embellishments. Salt and sweetening add flavor. Butter or oil improves the texture. With a good basic recipe you can achieve endless variations by adding different kinds of flours or grains, milk in place of water, eggs, herbs or other flavorings, or by varying proportions.

There are books listed in the bibliography which will help you find your way to all of the variations. *This book will provide you with a good, all-purpose bread recipe that could serve you forever if you didn't feel like* pursuing another one. It is not necessarily the greatest bread recipe in the world. It *is* a good one that is easy to make, and if you haven't baked bread before, it is a good way to begin.

At the risk of sounding mystical, I must tell you that one does not bake bread as simply as one boils an egg. The baking of bread is a ceremony. If it is approached and carried out as such, the baking will be not an ordinary act of cooking, but a deep, rich, satisfying, fulfilling experience.

The recipe as given calls for natural products only. If you don't have them or can't get them, substitute the products in parentheses instead. If you prefer white bread to whole wheat bread, use *un*bleached white flour. You may also substitute two or three cups of unbleached white flour for some of the whole wheat flour to make a somewhat lighter bread.

WHOLE WHEAT BREAD
    You will need (for two loaves):
2 packages of dry baking yeast
Lukewarm water
1 tablespoon sea salt (salt)
1/4 cup pure safflower oil (pure peanut oil, corn oil, melted butter)
1/4 cup organic honey (any pure honey)
8–10 cups stone-ground, whole wheat flour (whole wheat flour unbleached white flour)
1 egg
A large mixing bowl
2 bread pans
Cup, spoons, board, small bowl, towel

Pour two packages of dry baking yeast into a small bowl and add 1 cup of lukewarm water. The water should feel just comfortably warm to your finger. Stir and leave it for a few minutes until the yeast is dissolved.

In a large mixing bowl, combine

2-1/2 cups lukewarm water
1 tablespoon sea salt
1/4 cup oil
1/4 cup honey
8 cups flour
1 egg

Add the yeast mixture and stir the dough as well as you can with a large wooden spoon. When it is too difficult to mix, stop.

Now you are ready to knead. Sprinkle some flour on a large board or a clean table top. Spread the flour around and dust some on your hands. Scrape the dough out of the bowl and put it on the board. Dipping your hands in flour when necessary, flatten the lump enough so that you can fold it in half. Press down hard on the folded dough with the heels of your hands, first in one place, then in another. When the dough becomes quite flat, fold it in half again and knead some more. If you let your whole body rock forward as you knead, your arms won't get so tired. (They'll get somewhat tired no matter what you do.)

From time to time, turn the dough around. Keep folding and kneading. Flour your hands when necessary. After 5 or 10 minutes, you will begin to notice a

change in the texture of the dough. Less sticky, it will be firm enough so that it seems to "fight back" as you knead. When it doesn't cling much to your hands or the board, and when it is smooth and elastic and shiny on top, you have worked enough.

Without washing the large mixing bowl, rub the inside with oil or butter. Form the dough into a lump and put it into the bowl. Rub the top with a little oil to keep it from forming a hard crust as it dries out.

Cover the top of the bowl with a clean towel or cloth and set it in a warm place until the dough has risen to about twice its size. This takes about one hour. The tiny yeast plants feed on the sugars and starches in the dough and multiply rapidly. As they do, they give off carbon dioxide gas, and the gas bubbles puff up the dough. As soon as you put the bread into the oven, the heat will kill the yeast plants, and they won't grow any more.

When the dough has risen to double its bulk, as most recipes say, punch it down. That means just what you think it does. Make a fist and punch the dough a number of times until it goes down like a punctured balloon.

Knead the dough again (right in the bowl if you like) just for a minute or two, then divide it into two equal parts. Fold each piece over a couple of times and form it into the shape of a loaf. Slap it hard to break any big bubbles inside.

Put each loaf into an oiled or buttered loaf pan and shape it some more until it looks just the way you want it to. (If you don't have loaf pans, or if you want bread that looks different, bake it in oiled coffee cans. It will rise over the top and have a mushroom shape.) If you want your bread to have a soft crust, smear a little oil or butter on top of each loaf. If you prefer a crunchy crust, dip your fingers in cold water and rub it lightly over the tops of the loaves.

Cover the pans with a towel and let the dough rise once more until it has doubled in bulk, about 30–40 minutes.

Heat the oven to 350°F., put in the bread, and bake it for 50 minutes. As soon as you take the bread out of the oven, remove it from the pans and put it on wire racks to cool.

The aroma that fills your house is part of the ceremony. So is eating the first slice of bread, still warm from the oven, butter melting over the top and dripping down the sides.

If you don't have time to perform the bread-baking ceremony frequently, make four or six loaves at once and when they are thoroughly cool, wrap them well and freeze them. You might as well, because once you have been converted, you will never like bought bread again.

THE ORGANIC GOURMET

There is a popular notion (although its popularity is rapidly dwindling) that natural foods taste like straw, medicine, or worse. Just in case you come across an ignoramus who holds such an opinion, and you want to set him straight quickly, here are a few secret weapons guaranteed to do the job. While they all call for natural ingredients, it is of course desirable to use organic foods if possible. As usual, substitutions appear in parentheses.

YOGURT CREAM CHEESE

Take about a half yard of clean cheesecloth and tie up the four corners to make a bag with an open top. Pour homemade yogurt into the bag, tie a string around the top, and tie the string to the kitchen faucet. Let the liquid drain out of the yogurt into the sink for several hours or, in cool weather, overnight. Take the cheese out of the bag and season it with sea salt (or salt) and finely chopped chives, any other herbs such as garlic, tarragon, parsley, and so forth. You can turn it into a dessert spread instead by using a little bit of salt, honey, and chopped nuts to taste.

CREAM OF ZUCCHINI SOUP (COLD)

(or Cucumber or Watercress or Celery or Spinach or Anything Soup)

You will need:

1 onion
2 tablespoons butter
1/2 cup strong chicken broth
2 medium zucchini squash
1/2 cup milk
1 cup yogurt
Salt and pepper

Cut up a medium-sized onion and sauté it in a heavy saucepan in 2 tablespoons of butter. Do not let the onion brown. When it is yellow and tender, add the broth or strong chicken bouillon. Scrub the zucchini with a brush, but do not peel. Cut them into 1-inch slices and add them to the pan. Cover and boil about

5 minutes, until the squash is just beginning to get tender. Put into a blender and blend for a few seconds. Add the milk, yogurt, and salt and pepper to taste. Serve ice cold with chopped chives, a thin slice of lemon dusted with chopped chives or parsley, or a nasturtium blossom in each bowl.

SOYBEAN SNACKS

Put a cupful of rinsed soybeans into a bowl, cover them with water, and soak them in the refrigerator overnight. In the morning, drain the beans well and spread them out on a cookie sheet. Bake them in a 200°F. oven for 2-1/2 hours. Drizzle a teaspoonful of olive oil over the beans and stir them with a spoon until every bean looks oiled. Put them back into the oven for another half hour. As soon as you take them out, sprinkle them lightly with sea salt (or salt). When the beans are thoroughly cool, store them in a covered jar. Serve as you would salted nuts.

NUT PIE CRUST

You will need:

- 1/2 cup finely chopped unsalted walnuts, pecans, almonds, or other nuts
- 1 cup whole wheat (unbleached white) flour
- 5 tablespoons butter, room temperature
- 1/4 teaspoon sea salt (salt)

Mix the nuts, flour, and salt together. Blend in the butter with your fingertips. With your fingers, press the pastry into a 9-inch pie pan. Fill with any filling and bake until fruit filling bubbles or custard filling is set.

OPEN SESAME COOKIES

You will need:

1-1/2 sticks (3/4 cup) butter
1-1/2 cups brown sugar
1 egg
A few grains of salt
1 cup sesame seeds (finely chopped unsalted nuts)
1-1/4 cups whole wheat (unbleached white) flour
A cookie sheet

Melt the butter in a saucepan. Stir in the sugar, then add the egg and salt. Lastly, stir in the sesame seeds and flour. Mix until well blended. The mixture will never look completely smooth because of the seeds or nuts.

Butter a cookie sheet and place the mixture on it by the half-teaspoonful, leaving 2 inches between cookies. Flatten each one slightly with the bottom of a glass. Bake at 375°F. for about 10 minutes. The bottoms burn easily if left too long, so watch carefully. Remove the cookies from the pan with a spatula as soon as they are done. When they are thoroughly cool, store them in an airtight tin.

BANANA NUT CAKE

This is one of the most delicious desserts imaginable. If you bake only one cake in your whole life, make it this one!

You will need:

1 stick (1/2 cup) butter, room temperature
3/4 cup raw sugar (sugar)
1/2 cup honey
2 eggs
1 teaspoon baking soda
1/2 cup yogurt
3 average or 2 large ripe bananas mashed with a fork
1 teaspoon pure vanilla extract
1-1/2 cups unsifted whole wheat (unbleached white) flour
1 cup coarsely chopped or broken walnuts

Set the oven to 350°F. and butter a 9-inch tube pan or two loaf pans. Put the butter into a large mixing bowl and blend in the sugar. Beat in the eggs, then add all the other ingredients except the nuts. When the mixture is well blended (it will look lumpy because of the bananas), add the nuts. Put it into the pan (or pans) and bake for about 50 minutes or a trifle longer, *just* until the cake begins to shrink away from the sides of the pan. It should not be allowed to dry out in baking, but rather should have a moist, almost custardy texture. Serve as is or with a blob of whipped cream and a few banana slices. This cake will keep for days if it is well wrapped and stored in the refrigerator. It also freezes well.

GRANOLA (a superior breakfast cereal)
You will need:

4 cups oats
1 cup sesame seeds
1 cup cornmeal
1-1/2 cups whole wheat flour
1/2 to 1 cup chopped unsalted nuts
1 cup wheat germ
1/2 cup honey
1/2 cup pure oil
1 cup cold water

All ingredients should be organic if possible.

Mix everything together in a large bowl. Spread the mixture evenly in the bottom of a roasting pan. Bake in the oven at 250°F. for 2 hours, stirring well every 15 minutes. When the granola is cool, store it in jars and serve like any dry breakfast cereal, with milk and honey or raw sugar.

THE ORGANIC ECOLOGIST

As I begin to write this chapter, the front page of my daily newspaper carries a story datelined Houston, Texas. It reports that chemically laden smog felled over one hundred fifty persons in less than three months along the Houston ship channel. The channel, which is actually a canal, is lined along much of its fifty-mile length with heavy industry and is said to be one of the most polluted streams in the world. None of the one hundred fifty afflicted persons died, but almost all of them had to be hospitalized.

The article goes on to say that the Houston city administration has not been receptive to local groups seeking legal action against the industries. In its defense, the city has quoted federal studies to show that many other cities, including Los Angeles and New York, have even worse pollution problems than Houston.

There are similar shocking news stories almost daily. By now, there can hardly be anyone, anywhere, who does not know the importance of trying to restore our polluted and damaged earth. What to do and how to do it are huge and bewildering questions. Much has been written on the subject, but the mere prospect of reading —and carrying out—all the recommendations may be overwhelming.

If you find it difficult to change your entire mode of thinking and living all at once, it may make it easier for you to get started if you try to follow just one basic rule: *Remember your own role in the order of life on earth. Do not separate yourself from the rest of nature. Know that everything you do, or fail to do, affects every living thing.*

Once you have thoroughly absorbed that idea and made it part of yourself, you can begin to think of specific kinds of actions. Here are a few thoughts to help stimulate your own.

Don't pollute. Just as you care for your interior ecology by not eating pollutants, care for the ecology of the earth by the following measures.

1. Don't use poison sprays or pesticides.
2. Use only soaps, not detergents, for laundry or cleaning.
3. Use pure soaps for washing yourself, too. Avoid hexachlorophene, which kills desirable bacteria along with the unwanted kind.

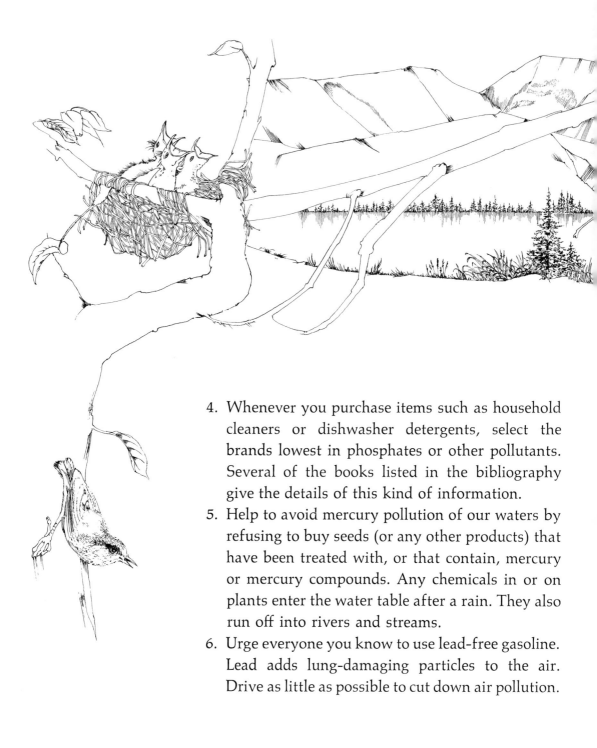

4. Whenever you purchase items such as household cleaners or dishwasher detergents, select the brands lowest in phosphates or other pollutants. Several of the books listed in the bibliography give the details of this kind of information.
5. Help to avoid mercury pollution of our waters by refusing to buy seeds (or any other products) that have been treated with, or that contain, mercury or mercury compounds. Any chemicals in or on plants enter the water table after a rain. They also run off into rivers and streams.
6. Urge everyone you know to use lead-free gasoline. Lead adds lung-damaging particles to the air. Drive as little as possible to cut down air pollution.

Form car pools. Use a bicycle and public transportation. Learn to walk again.
7. Don't burn leaves. If your community doesn't have a removal program, agitate for one. Fire consumes oxygen, and even good organic material makes black smoke.
8. Avoid disposable plastic. Whenever you have a choice, buy liquids in glass rather than plastic containers. Use plastic bags and wraps as little as possible; when you can, wash and reuse them. Most city garbage is incinerated, and when plastics are burned, they give off noxious, polluting gases. Plastic products that are not burned never decay as paper does, but remain as litter for years.
9. Don't *ever* litter. Try to organize a community campaign to make people aware of the pollution of the earth by littering. Start a clean-up action.

Conserve. We are a nation of outrageous users and wasters. Buy only what you really need. Make it, grow it, cook it, bake it. Refuse to buy items that have no real use. Refuse unnecessary wrappings or packing. Use only materials that can be used over and over again, recycled. Some specific measures:

1. Refuse to buy nonreturnable bottles. Tell your storekeeper. Support community action to collect bottles from people who are too lazy to turn them in. In some areas, glass is collected for recycling.
2. Save paper. Our trees and forests are wasted to make paper that we can do without. By-products of paper-making pollute our rivers. Don't throw away newspapers; recycle them. Call your local junk dealer. Use cloth napkins, handkerchiefs, and diapers instead of paper ones. Buy the no-iron kinds, and it won't be much trouble. Take a market basket or shopping bag to the store with you and refuse paper bags. Save gift-wrapping and use it again. When you do buy paper products, buy white instead of colored ones. The dyes pollute.
3. Avoid buying cans when possible. If you do, buy aluminum cans only and turn them back to the aluminum companies for recycling. Help your community establish an aluminum collection station to raise money for a local charity while it conserves.
4. Save water. Good water is a scarce natural resource in many areas, becoming scarce in others.

Repair leaky faucets or drips of any kind. Don't run water needlessly. Put a brick in your toilet tank to cut down the amount of water it uses. Take a shower instead of a bath.

5. Save electricity and all other forms of energy. The production of electricity, gas, or coal involves the burning of natural fuels, polluting of air and water, destroying land with strip mines, and overheating waterways. Use as few appliances as possible. Use fluorescent lamps instead of incandescent ones when feasible. Turn off lights and appliances when you don't need them.

6. Help to organize local garbage recycling or a community compost heap. The village of Scarsdale, New York, invested $17,500 in leaf-gathering equipment. Villagers pile leaves in front of their houses, and the leaves are collected once a week for a two-month period each fall. The leaves are then hauled to a four-acre area where they are used as the green matter in a number of compost heaps. A year later the compost is sold to local citizens on a come-and-get-it basis for six dollars a cubic yard. As the program grows, Scarsdale hopes to produce enough compost to sell it commercially. Even in its first year of operation, the plan saved the village an estimated $58,000 compared to the cost of conventional methods of leaf disposal. But, of course, the dollar saving is far less important than the conservation and ecology values of such a community project.

7. Consume less. Help to make it fashionable in your

community *not* to have the latest model car, the most up-to-date house, the newest style in clothing. Many of the world's people consider Americans foolish in their ostentatious use of consumer goods. Europeans pride themselves on building cars that last for a long time, that don't change their styling every year, and that use little fuel. Europeans waste less than Americans in general, although, now that they are enjoying a higher standard of living than ever, the richer countries of the world are beginning to have some of the same problems we are.

Neither Americans, Europeans, nor Asians can go on as they are much longer. We have littered our countryside with junk. We have poisoned our air and water, depleted our forests, killed our wildlife. We may well kill ourselves if we don't reverse the trend.

Ecology and conservation are part of organic living. Do the best you can personally. Influence those around you. Start action committees in your neighborhood school, church, club. Make your ideas known to local shopkeepers and to manufacturers. Don't think it doesn't help.

On the same day that my newspaper reported the smog in Houston, it reported another development. A number of chemical companies have stopped manufacturing pesticides. The article said that although most of the companies claimed their reasons were economic, some admitted "privately" that they were having trouble with "ecological critics."

What is an ecological critic? An ecological critic is someone like you. It is someone who believes in organic living and someone who believes in it strongly enough to try to save our earth.

FOR FURTHER READING

BECK, Bodog F., and SMEDLEY, Dorée. *Honey and Your Health*, paper ed. New York: Bantam Books, 1971.

BROWN, Edward E. *The Tassajara Bread Book*, paper ed. Berkeley, Calif.: Shambala Publications, 1970.

CAILLIET, Greg M., SETZER, Paulette Y., and LOVE, Milton S. *Everyman's Guide to Ecological Living*, paper ed. New York: The Macmillan Company, 1971.

CARSON, Rachel. *Silent Spring*, rev. paper ed. New York: Fawcett World Library, 1970.

DARLINGTON, Jeanie. *Grow Your Own*, paper ed. Berkeley, Calif.: The Bookworks, 1970.

DAVIS, Adelle. *Let's Cook It Right*, paper ed. New York: New American Library, 1970.

DAVISON, Verne E. *Attracting Birds: from the Prairies to the Atlantic*. New York: Thomas Y. Crowell Company, 1967.

FRAZER, Sir James G. *The Golden Bough*. New York: The Macmillan Company, 1951.

GIBBONS, Euell. *Stalking the Good Life*, paper ed. New York: David McKay, 1971.

GOLDSTEIN, Jerome, and GOLDMAN, M. C. *Guide to Organic Foods Shopping and Organic Living*, paper ed. Emmaus, Pa.: Rodale Press, 1970.

GRAHAM, Frank, Jr. *Since Silent Spring*, paper ed. New York: Fawcett World Library, 1970.

HUNTER, Beatrice Trum. *Gardening Without Poisons,* paper ed. New York: Berkley Publishing Corporation, 1971.

———. *The Natural Foods Cookbook,* paper ed. New York: Simon and Schuster, 1961.

KOHN, Bernice. *The Beachcomber's Book.* New York: The Viking Press, 1970.

KORDEL, Lelord. *Cook Right—Live Longer,* paper ed. New York: Award Books, 1966.

LAUREL, Alicia Bay. *Living on the Earth,* paper ed. New York: Vintage Books, 1970.

MITCHELL, John G., and STALLINGS, Constance L., eds. *Ecostatics: The Sierra Club Handbook for Environmental Activists,* paper ed. New York: Pocket Books, 1970.

ORGANIC GARDENING. *Guide to Organic Living,* paper ed. Emmaus, Pa.: Rodale Press, 1971.

PENDERGAST, Chuck. *Introduction to Organic Gardening,* paper ed. Los Angeles: Nash Publishing, 1971.

RODALE, Robert. *The Basic Book of Organic Gardening,* paper ed. New York: Ballantine Books, 1971.

Ross, Shirley "Wonderful." *The Interior Ecology Cookbook,* paper ed. San Francisco: Straight Arrow, 1970.

SIMON, Seymour. *Discovering What Earthworms Do.* New York: McGraw-Hill Book Company, 1969.

SMETINOFF, Olga. *The Yogurt Cookbook,* paper ed. New York: Frederick Fell, 1966.

SWATEK, Paul. *The User's Guide to the Protection of the Environment,* paper ed. New York: Ballantine Books, 1970.

TEICHNER, Mike and Olga. *The Gourmet Health Foods Cookbook,* paper ed. New York: Paperback Library, 1967.

TERRES, John K. *Songbirds in Your Garden.* New York: Thomas Y. Crowell, 1968.

TURNER, James S. *The Chemical Feast,* paper ed. New York: Grossman Publishers, 1970.

U.S. DEPARTMENT OF AGRICULTURE, *Minigardens for Vegetables,* Home and Garden Bulletin No. 163. Washington, D.C.: U. S. Government Printing Office, 1969.

VAN BRUNT, Elizabeth R. *Handbook on Herbs,* paper ed. New York: Brooklyn Botanic Garden, 1958.

VINE, Lesley. *Ecological Eating,* paper ed. New York: Tower Publications, 1971.

WADE, Carlson. *Health Food Recipes for Gourmet Cooking.* New York: ARC Books, 1969.

PERIODICALS

National Wildlife. National Wildlife Federation, 381 West Center Street, Marion, Ohio 43302.

Natural Life Styles. P.O. Box 150, New Paltz, N.Y. 12561.

Organic Farming and Gardening. Rodale Press, 33 East Minor Street, Emmaus, Pa. 18049.

Ranger Rick's Nature Magazine (for children). National Wildlife Federation, 381 West Center Street, Marion, Ohio 43302.

Smithsonian. Smithsonian Associates, 900 Jefferson Drive, Washington, D.C. 20560.

INDEX

Additives, 14–15, 17
Alfalfa-seed sprouts, 52, 54
Aluminum cans, recycling of, 80
Artificial colors and flavors, avoidance of, 14

Bacteria, in compost heap, 32
Banana nut cake, recipe for, 74
Bean sprouts, 51, 52
Beans, 58
Beetles, tiger, for pest control, 49
Berries, wild, 59
Blackberries, 59
Bleached white flour, 16, 18
Blueberries, 59
Bottles, nonreturnable, refusal to buy, 80
Bran, 57
Bread: homemade, 62–69; manufacturers' additives in, 14; sacramental use of, 63, 64

Buckwheat, 57
Bulgar, 57
Bulgarian yogurt culture, 26
Butter, 57
Buttermilk, 58

Cake, additives in, 14
Cancer, 58
Carson, Rachel, 46, 47
Cereal: additives in, 14; whole grain, 57
Cheese, 58; processed, 14, 58
Cholesterol, 18
Clams, 59
Colander, for growing sprouts, 52, 53
Companion planting, 50
Compost, 29, 30–34, 44, 81
Conservation, 10, 80–83
Corn oil, 57
Cornmeal, 57
Cottonseed oil, 57
Crabs, 59

Damselflies, for pest control, 49
DDT, 47
Detergents, avoidance of, 78

Earthworms, usefulness of, 28
Ecological critic, 83
Ecology, 10, 28, 34, 46, 47, 61, 76–83
Eggs, fertilized, 58
Electricity, saving, 81
Energy, saving, 81

Fats: for cooking, 57; hydrogenated, 18, 57, 58
Fertilizer, natural, 29
Fish, 59; mercury in, 13
Flour, whole grain, *see* Whole grain flour
Fluorescent lamps, use of, 81
Foods: additives in, 14–15, 17; back-to-nature, 56–61; refined, 16; wild, 58–61
Frankfurters, additives in, 14
Frazer, James, 63
Fruit drinks, benzoate of soda in, 15
Fruit juices, unsweetened, 58
Fruits, unsulphured and unfumigated, 58

Garden: gardenless, 37–45; organic, 27–36
Gasoline, lead-free, 79
Gelatin, 58
Golden Bough, The (Frazer), 63
Gourmet, organic, 70–75
Grains, whole, 16, 18, 57

Granola, recipe for, 75
Green matter, in compost, 31, 32, 33

Herbs, 45, 58
Hexachlorophene, avoidance of, 78
Honey, 16, 18, 20, 57, 61
Hormones, in meats, 58
Howard, Albert, 30
Humus, 31, 33, 34
Hydrogenated fats, 18, 57, 58

Indore composting method, 30–34

Ladybirds, for pest control, 28, 49
Littering, avoidance of, 79

Manure, 31, 32, 33
Maple syrup, 57
Maraschino cherries, benzoate of soda in, 15
Margarine, 57
Matzoh, 64
Meats: additives in, 14; without hormones, 58
Mercury pollution, 13, 78
Milk, raw certified, 58
Millet, 57, 63
Minerals, 16, 52
Molasses, unsulphured, 57
Mulberries, 59
Mulch, 28, 34–35
Mung bean sprouts, 52, 54
Mushrooms, poisonous, 61
Mussels, 59

Nut pie crust, recipe for, 72

Nuts, 58

Oats, 57
Oils, for cooking, 57
Olive oil, 57
Organic garden, 27–36

Paper, saving, 80
Pea sprouts, 52, 54
Peanut butter, 18
Peanut oil, 57
Pest control, in organic garden, 28, 36, 47–50
Pesticides, 46, 78, 83
Plastics, 79
Pollution control, 10, 77–79
Praying mantises, for pest control, 28, 49
Proteins, 52
Purslane, 59

Raspberries, 59
Recycling: of aluminum cans, 80; with compost, 29, 81; of garbage, 81; glass collected for, 80
Rice: brown, 16, 57; white, 16, 18
Rye, 18, 57

Safflower oil, 57
Sassafras, 59–61
Scarsdale (N.Y.), leaf disposal in, 81
Seaweed, 59
Seeds, planting, 40, 41, 42
Sesame cookies, recipe for, 73
Sheep sorrel, 59
Sieve, for growing sprouts, 52, 53

Silent Spring (Carson), 46, 47
Smog, 76, 83
Soft drinks, benzoate of soda in, 15
Soybean grits, 57
Soybean oil, 57
Soybean snacks, 72
Soybean sprouts, 54
Sprouts, 51–55
Strawberries, 59
Sugar: raw, 16, 18, 57–58; white, 18, 58
Sunflower oil, 57

Tiger beetles, for pest control, 49
Top dressing, in organic garden, 34

Unbleached white flour, 16, 65

Vermiculite, 40
Vitamins, 16, 52, 57

Wasps, trichogramma, for pest control, 49
Water, conservation of, 80–81
Weeding, 44
Whole grain flour, 16, 18, 57; stone-ground, 57
Whole wheat bread, homemade, 65–69
Wildlife, 83

Yogurt, 19–26, 58; recipe for, 22; value of, 20
Yogurt cream cheese, 71

Zucchini soup, recipe for, 71–72

ABOUT THE AUTHOR

BERNICE KOHN was born in Philadelphia. She entered the University of Wisconsin as a creative writing major, but soon found herself drawn to science. The combination of these two interests has resulted in more than three dozen children's books. Her most recent, for Viking, is *The Beachcomber's Book.*

Bernice Kohn is the mother of three grown children. She has been reading labels, baking bread, and gardening organically for many years. She and her husband, the author Morton Hunt, divide their time between New York City and East Hampton, New York.